# iPhone 16 Pro Made Easy

The Ultimate Step-by-Step Guide for Seniors & Beginners to Master Hidden Features, Smart Tips, and Apple Intelligence

Blake Orion

# Contents

# INTRODUCTION

## Welcome to the iPhone 16 Pro Made Easy Journey

If you've ever picked up your iPhone and felt overwhelmed, frustrated, or left behind—this book was written for you.

The iPhone 16 Pro is one of the most powerful, beautifully designed smartphones ever created. It's fast, intelligent, and full of features that can simplify your life, entertain you, keep you safe, and even help you reconnect with the people you love. But here's the truth: it can also feel intimidating.

Apple assumes that everyone who buys their devices is a tech genius. They expect you to "just know" how to use gestures, customize settings, use Siri, explore hidden features, and manage privacy like a digital ninja. But what about the rest of us?

What about seniors who didn't grow up with smartphones? Or busy parents and professionals who don't have time to figure everything out on their own? Or beginners who just want a friendly, step-by-step guide without the tech jargon?

**That's exactly why this book exists.**

This guide was built for real people—people like you—who want to get more out of their iPhone without the headache. Whether you're unboxing your first iPhone today or upgrading from an older model, you'll find clear, simple instructions that walk you through everything—from setup to hidden features you never knew existed.

Here's what makes this book different from every other manual out there:

- **We keep it simple**: No technical mumbo jumbo. Just real-world explanations with real-life examples.
- **We show, not just tell**: Every major step includes visual guidance and screenshots so you never feel lost.
- **We focus on you.** Whether you're 70 or 17, tech-shy or curious, we'll meet you where you are and walk with you.
- **We include the "why," not just the "how"**: It's not just about knowing what to do—it's about understanding why it matters.

You'll learn how to use your iPhone 16 Pro to take stunning photos, keep your personal data private, stay in touch with family, navigate with ease, and discover tips that even seasoned iPhone users miss. You'll even get insider tricks on how to save battery, use Apple's AI-powered features, and troubleshoot common problems.

Most importantly, by the time you finish this book, you'll be *in control* of your iPhone—not the other way around.

So if you've ever said…

- "I just want to use my phone without being confused."

- "Why does everyone else make it look so easy?"
- "I don't want to call my kids or grandkids every time I get stuck."

...then you're in the right place.

Turn the page. Let's take this journey together. Your iPhone is about to go from frustrating to fantastic—and you'll finally enjoy everything it was designed to do for you.

# Chapter 1

## Welcome to Your iPhone 16 Pro

### Discovering the Power in Your Hands

If you're holding the iPhone 16 Pro for the first time, take a moment to appreciate what you've got in your hands—it's more than a phone. It's a camera, a notebook, a health tracker, a photo album, a smart assistant, and your personal connection to the world.

But if it feels unfamiliar or even a little intimidating, don't worry. You're not alone—and you don't have to figure it out by yourself. This chapter is your calm, friendly introduction to everything that makes the iPhone 16 Pro special.

### What's New in the iPhone 16 Pro?

Before we dive into how to use your phone, let's look at what makes the **iPhone 16 Pro** different from previous iPhones—or any smartphone, really. Apple's latest release includes a combination of powerful features that are easy to appreciate once you understand them:

- **Apple Intelligence (AI) Integration**: Your iPhone can now understand context, make suggestions, and complete tasks for you, almost like magic.

- **A17 Bionic Chip**: It's the brain behind the speed—making apps load faster, photos process quicker, and everything feel seamless.
- **Advanced Camera System**: With enhanced night mode, 5x optical zoom, and cinematic video, it's like carrying a pro-grade camera in your pocket.
- **Longer Battery Life**: More time for calls, messages, and browsing—less time hunting for a charger.
- **Customizable Action Button**: Replace the silent switch with a shortcut you actually use, like launching the camera or activating a voice memo.
- **Enhanced Privacy Tools**: Greater control over what data you share, and with whom.

Whether this is your first iPhone—or your tenth—this version has more depth, intelligence, and customization than ever before.

## iPhone 16 Pro vs. Pro Max vs. Regular iPhone 16

If you're wondering what the difference is between the models, here's a quick side-by-side to help:

Feature iPhone 16 16 Pro iPhone 16 Pro Max **Display Size** 6.1 inches6.1 inches (ProMotion), 6.7 inches (ProMotion) Material Aluminum + glass, Titanium + Ceramic Shield Titanium + Ceramic Shield **Camera Capabilities** Dual-camera system, triple-lens system w/ LiDAR, 5x telephoto, triple-lens Life: good, better, best. Price: lower, mid-range premium.

If portability is important, the **iPhone 16 Pro** gives you the best of both worlds—premium performance without the bulk of the Max model.

## What's in the Box?

Inside your iPhone 16 Pro box, you'll find:

- The iPhone 16 Pro
- USB-C charging cable
- Quick Start guide
- SIM removal tool (for some regions)

*Note: Apple no longer includes a charging adapter. If you don't already have a USB-C charger, you'll need to purchase one separately.*

## A Quick Tour of the Device

Let's get familiar with your phone's exterior. Don't worry—you'll never have to remember it all at once. We'll revisit each button as you need it.

### Front:

- **Display**: Where all the action happens
- **Dynamic Island**: The notch at the top—now used for alerts, timers, music, and more

### Left Side:

- **Volume Up/Down Buttons**: Control volume, take photos, and more
- **Action Button** (replaces the silent switch): Customize it for camera, flashlight, voice memos, or any shortcut

### Right Side:

- **Power Button**: Press to wake, hold for Siri or shutdown options

### Bottom:

- **USB-C Charging Port**
- **Speakers & Microphones**

### Back:

- **Triple-Lens Camera System**
- **LiDAR Scanner** (helps with photos and AR experiences)

## You're in Good Hands

By the end of this book, the iPhone 16 Pro won't just be a smartphone—it will feel like a trusted companion. Each chapter is designed to build your confidence, at your pace, with real-life examples and screenshots along the way.

# Chapter 2

## Unboxing & Setup for Non-Techies

### A Gentle, Confident Start to a Smarter Digital Life

Take a deep breath.

You don't need to be a tech wizard, digital native, or software engineer to set up your iPhone 16 Pro. You only need one thing: the **willingness to start**—and this chapter will take care of the rest.

Setting up your iPhone isn't just about getting a device to work. It's the first step in crafting a tool that's **uniquely yours**—designed to make your days easier, more joyful, and more connected.

Whether this is your very first iPhone or your fifth upgrade, we're going to move step-by-step in a calm, friendly, and clear way. You'll finish this chapter not just with a phone that's ready to go, but with a sense of confidence and accomplishment.

### What's in the Box: Understanding Each Item

Let's begin with what you'll find inside that sleek white box.

- **Your iPhone 16 Pro**
  Wrapped in protective film, this is your new smart companion.

- **USB-C Charging Cable**
  This charges your phone and can be used for data transfer.

- **SIM Ejector Tool** *(in some regions)*
  Helps insert or remove your SIM card if you're using a physical SIM.

- **Quick Start Card**
  A mini welcome guide from Apple (but we'll do a better job here!)

## Mindset First: You Are Not Behind

Before we even turn the device on, here's something important to remember:

**Technology isn't just for the young. It's for the curious.**
No one is "too old" or "too late." You're exactly where you need to be.

Every step in this book is written with the understanding that you might feel hesitant—and that's okay. It's not your job to *already know* how things work. That's what this guide is for.

## Step 1: Powering On and the First Hello

1. Press and hold the **power button** (on the right side) until the Apple logo appears.

2. Wait for the **"Hello"** screen to show in multiple languages.

You've just awakened your device for the first time. That glowing screen? That's your blank canvas.

## Step 2: Language, Country & Wi-Fi

1. **Select your language** (e.g., English)
2. Choose your **country or region**
3. Connect to your **Wi-Fi network**
   - Select your home network and enter your password (you can find this on your router or from your internet provider).
   - Tap "Next" and wait a moment.

Why Wi-Fi? Wi-Fi connects your iPhone to the internet for updates, downloads, and andcommunication. You can always switch to mobile data later if needed.

## Step 3: Setting Up Face ID or Passcode

Your phone will ask if you'd like to use **Face ID**—a secure way to unlock your phone using your face.

- You can **set it up now** (just follow the on-screen instructions to move your head in a circle).
- Or **skip it and do it later**—either way, your phone will still work just fine.

After this, you'll be asked to **create a passcode** (a 6-digit code to protect your data).
Make sure to write it down somewhere safe—you'll need this often.

*Tip:* If you ever forget your yourpasscode, don't panic. There's a recovery process, and we'll cover that later in the book.

## Step 4: Apple ID — Your Digital Passport

The Apple ID is your key to everything Apple-related: downloading apps, syncing photos, backing up your data, and more.

- If you have an Apple ID from a previous device, enter it now.
- If you don't, tap **"Create a New Apple ID"** and follow the steps (name, birthday, email, password).

*Don'twant to use your real email?* You can create a free iCloud email as part of thisprocess (e.g., yourname@icloud.com).

This step might take a few minutes—but it's essential. Think of your Apple ID like a **digital vault key** to everything you'll store or access through your iPhone

## Step 5: Restore, Transfer, or Set Up as New?

Apple gives you a choice:

- **Restore from iCloud Backup** (if you've used an iPhone before)
- **Transfer from Android** (if switching from another phone brand)
- **Set Up as New iPhone** (if this is your first time or you

want a clean slate)

We recommend **"Set Up as New"** if you're a beginner—it's the easiest, and we'll guide you every step of the way.

## Step 6: Enabling Apple Features (You Can Say "No")

Next, your iPhone will walk you through a few options:

- Enable **Location Services** (helps with maps, weather, Find My iPhone)
- Enable **Siri** (Apple's voice assistant)
- Set up **Screen Time**, **Automatic Updates**, etc.

You can choose to say **"Set Up Later"** for anything you're unsure about. Nothing here is permanent.

> *FriendlyTip:* You're in control. You can always turn features on or off later in **Settings.**

## Step 7: Your Home Screen Awaits

That's it! Your phone is now set up and ready.

You'll land on the **Home Screen**—where all your apps live. From here, you can make calls, send messages, take photos, explore the web, and so much more.

## A Personal Note Before You Continue

It might have felt like a lot of steps—but you did it.

You now hold a powerful, intelligent, and highly personalized device in your hand. Not because you're tech-savvy (yet)—but because you were willing to learn.

**That makes you smarter than the phone itself.**

# Chapter 3

## The Home Screen Decoded

### Learning to Navigate with Confidence and Ease

Think of your home screen like the front door to your digital home.

When you unlock your iPhone, this is the first thing you see. It's where your favorite apps live, where new alerts pop up, and where you launch everything from phone calls to photos. In this chapter, we'll take a slow, thoughtful walk through each part of your Home Screen—so you feel in full control from the moment you wake your phone.

By the end of this chapter, you'll no longer feel like a visitor in someone else's digital house. You'll feel at home.

### Orientation: What You're Looking At

Here's what you're seeing when you first unlock your iPhone:

- **Top Area (Status Bar):**
    - Shows time, signal strength, Wi-Fi, and battery.
    - Some indicators (like Do Not Disturb) may appear briefly here.

- **Dynamic Island (Top Center):**
    - This is Apple's newest feature—like a tiny window into what your phone is doing.
    - It displays ongoing activities (timers, music playing, FaceTime calls) in real time.
- **App Icons (Main Area):**
    - These are shortcuts to your apps.
    - Tap any icon once to open it. Press and hold to move or delete it.
- **Dock (Bottom Row):**
    - Always visible as you swipe between screens.
    - Holds your most-used apps—like Phone, Safari, Messages, and Music.

## Gestures That Make Life Easier

With the removal of the Home button in newer iPhones, Apple introduced **gesture navigation**. Don't worry—it's simpler than it sounds. Think of gestures like "swiping" motions that help you move between apps and actions.

Here are the essentials:

- **Swipe Up from Bottom (Anywhere):**
  Go back to the Home Screen.

- **Swipe Up & Hold:**
  Opens the **App Switcher** (lets you see and switch between open apps).

- **Swipe Down from Top Right:**
  Opens the **Control Center** (quick access to brightness, volume, Wi-Fi, etc.).

- **Swipe Down from Top Left or Center:**
  Opens **Notifications**—missed calls, messages, app alerts.

- **Swipe Right on Home Screen:**
  Opens the **Today View** with widgets and Siri Suggestions.

## Organizing Your Apps

If your apps feel scattered or overwhelming, let's fix that.

### Move Apps:

1. Press and hold an app icon until they all start to jiggle.
2. Drag the app where you want it.
3. Tap **Done** (top right) when you're finished.

### Create a Folder:

1. Drag one app on top of another—it automatically creates a folder.
2. Rename the folder (e.g., "Health" or "Social").

3. Add more apps by dragging them in.

### Add to Dock:

1. Remove an app from the dock by dragging it upward.
2. Drag your new favorite app into the now-empty space.

ProTip: Keep only your **most-used apps on the first page.** You can swipe left for more.

## Widgets: Your Phone, at a Glance

Widgets are live tiles that show **real-time info** without having to open the app. Think of weather, calendar, step count, photos, or news.

### Add a Widget:

1. Long-press the Home Screen, then tap the **+ icon** in the top left.
2. Choose a widget (like Clock or Fitness).
3. Tap "Add Widget" and drag it where you want.

## App Library: Your Built-In Organizer

Swipe all the way to the right, and you'll enter the **App Library**. This is where **every single app** on your phone is automatically grouped and stored.

- You don't need to organize this; it updates on its own.
- Use the **search bar at the top** to find any app instantly.

## You're In Charge Now

You've just taken a powerful step forward. The home screen is no longer a mystery—it's a tool. A dashboard. A canvas for you to shape the way you live, connect, and play.

From this point forward, you'll begin customizing your iPhone to reflect **your life, your needs, and your rhythm.**

# Chapter 4

## Mastering Apple Intelligence

### Turning Your iPhone into a Truly Smart Companion

Imagine if your phone could anticipate your needs. Remind you of what matters before you even ask. Suggest the right words, open the right apps, and save you time with just a glance or a tap.

That's not science fiction anymore. It's Apple Intelligence—built directly into your iPhone 16 Pro.

In this chapter, you'll learn how Apple's new AI-powered features can simplify your life, streamline everyday tasks, and make your iPhone feel less like a gadget and more like a helpful assistant.

And don't worry: you don't need to understand "artificial intelligence" or tech terms. Everything here is real, practical, and tailored to *you*.

### What Is Apple Intelligence?

Apple Intelligence is a suite of smart tools baked into your iPhone that helps your phone understand your habits, preferences, and needs.

It's like giving your phone a brain—and a heart.

Apple Intelligence doesn't just react—it anticipates. And it's designed with your **privacy** and **simplicity** in mind.

## Everyday Things Apple Intelligence Can Do

Here are some practical ways Apple Intelligence can help you from the moment you wake up to the time you go to bed:

- Suggest reminders based on your conversations or emails.
- Offer quick replies to texts that sound like *you*.
- Clean up or summarize long emails or notes in seconds.
- Automatically sort photos and documents based on your past behavior.
- Create calendar events just by reading your messages.
- Offer app suggestions based on time, place, or routine.
- Help you write better messages, journal entries, or even social media posts.

You don't have to do anything complicated. Just use your phone like normal, and over time, it begins to learn what makes sense for *you*.

## Setting Up Apple Intelligence Features

You don't have to "turn on" Apple Intelligence with a switch—it runs quietly in the background once your phone is updated and connected to your Apple ID. But there are a few smart features you'll want to check and customize.

## Smart Suggestions

Your phone may suggest

- Apps you typically use at certain times (e.g., music in the morning).
- Contacts you message most often when opening Messages.
- Actions like starting a workout or launching Maps when you're at the gym or in the car.

These appear automatically on your Lock Screen, in Spotlight Search, or as Siri Suggestions. You can fine-tune these in

**Settings > Siri & Search > Suggestions**

From here, you can turn suggestions on or off for specific apps.

## Using Siri with Apple Intelligence

Siri is smarter than ever—and it works hand-in-hand with Apple Intelligence.

You can now say:

- “Siri, summarize this article for me.”
- “Siri, what did Mom text me yesterday?”
- “Siri, remind me to take my medication when I get home.”
- “Siri, write a polite reply to this message.”

What's different now is Siri's ability to understand *context*. If you're looking at a photo, email, or article, you can refer to it directly—like saying, "Add this photo to my journal" or "Forward this to Anna."

To get the most out of Siri:

1. Go to **Settings > Siri & Search**
2. Turn on **Listen for 'Hey Siri'** and **Press Side Button for Siri**
3. Enable **Allow Siri When Locked** if you want hands-free access anytime

## Communication Made Easier

Apple Intelligence makes communicating smoother and more natural:

- **Smart Replies**: When you get a message, your phone may suggest quick replies you can send with one tap.
- **Writing Assistance**: Whether you're texting, emailing, or writing a note, you can get suggestions for better tone, grammar, or clarity.

For example, if you type a message and hold down on it, you might see options like "Make it more polite" or "Make it more concise."

This is especially helpful when writing to doctors, coworkers, or customer service.

## Smarter Photos and Visual Understanding

Apple Intelligence doesn't just read—it sees.

Your iPhone can now:

- Recognize pets, people, places, and text in photos
- Search for "photos of my granddaughter in the park."
- Automatically organize images into collections
- Lift text out of any photo (like a phone number on a sign)

Just open the **Photos app**, tap **Search**, and try typing something like

- "Screenshots from April"
- "Pictures with Dad"
- "Recipes I saved last week"

You can even copy text from images by pressing on it—just like it's normal text.

## Journaling with Intelligence

If you're using Apple's new **Journal** app, you'll notice how Apple Intelligence quietly supports reflection and well-being.

It may suggest moments to remember—like a great photo, a meaningful conversation, or a completed fitness goal. You can choose to write about them, save them, or create entries with just a few taps.

This is optional but powerful. It helps you capture life's small wins and beautiful memories—even if you forget to write them down yourself.

## Your Privacy Is Respected

Apple has designed its AI tools differently from other companies. Your personal information **stays on your device**—not on remote servers.

That means

- Apple doesn't read your messages, photos, or contacts.
- Data is processed on your iPhone (or Apple's private cloud if needed—but never shared or sold).
- You're in complete control of what gets shared.

It's intelligence without invasion. Smart help without sacrificing your privacy.

## You Don't Need to Master It All at Once

One of the most powerful things about Apple Intelligence is how quietly it works. You don't need to learn it all now. You don't need to activate everything. You don't need to remember commands.

As you continue using your iPhone, it will slowly adjust to your patterns and offer help in small, meaningful ways. You'll simply notice that life feels a bit more organized. A bit easier. A bit more "tuned in."

And when it does, you'll know that you're not just using a phone anymore—you're *collaborating with it.*

# Chapter 5

## Taking Incredible Photos and Videos

### Capture Your Life Like a Pro—No Experience Needed

There's a moment that happens when someone uses the iPhone camera for the first time and sees what it can do. The colors look richer. The people look more vibrant. And for many, the response is the same:

*"Wow… I actually took that?"*

The iPhone 16 Pro camera is more than a lens. It's a memory-maker, a time-freezer, a beauty-capturer. And the best part? You don't need to be a photographer to use it.

This chapter will help you understand your camera, unlock its features, and start taking photos and videos that make people stop and smile.

## Why the iPhone 16 Pro Camera Stands Out

The camera on your iPhone 16 Pro is more advanced than many standalone digital cameras. It includes:

- A **triple-lens system** for wide, ultra-wide, and telephoto shots

- **Night mode** for low-light magic
- **Macro mode** for stunning close-ups
- **Cinematic video** for movie-like clips
- **Smart HDR and AI enhancements** that auto-improve your shots

All of that might sound technical. But here's the truth:

**The iPhone does most of the hard work. You just point and tap.**

## How to Open the Camera—Fast

- From the **Lock Screen**: Swipe left instantly.
- From the **Home Screen**: Tap the **Camera app**.
- With your **Action Button** (**if** customized for camera access):
- Ask Siri, "Open the Camera."

Don't worry—you can open it a dozen times a day. It's fast, safe, and always ready.

## Understanding the Camera Modes

When you open the Camera app, you'll see a carousel of modes at the bottom. Let's walk through the ones you'll use most often.

## Photo (Default)

- Ideal for most everyday shots.
- Point your phone at what you want to capture.
- Tap the shutter button (white circle) once.

## Portrait

- Automatically focuses on a person or object, blurring the background slightly for a professional look.
- Perfect for close-ups of people, pets, flowers, or keepsakes.
- The camera will guide you—just follow its instructions like "Move closer" or "Hold still."

## Night Mode

- Activates automatically in dark environments.
- Hold the phone steady, and the camera will brighten the image using multiple frames.
- A countdown timer may appear—don't move until it finishes.

## Video

- Swipe to "Video," then tap the red button to start recording.
- Tap again to stop.

- Works great for events, family moments, or scenic views.

### Cinematic

- Blurs the background while keeping the subject in sharp focus—like a movie.
- Automatically tracks people as they move.
- Perfect for interviews, storytelling, or kids playing.

### Slow-Mo and Time-Lapse

- Slow-mo: Great for capturing water, action, or dance.
- Time-lapse: Use this when filming clouds, city scenes, or long moments in short clips.

> Tip: You don't need to remember every mode. Start with **Photo** and **Video**—you can explore the rest as you gain confidence

### Zooming In and Out

On the screen, you'll see buttons labeled. **5x**, **1x**, **2x**, and possibly **5x** depending on your model.

- **5x** = Ultra-Wide: Fits more in the frame—great for landscapes or group shots.
- **1x** = Standard: Everyday photos.
- **2x or 5x** = Telephoto Zoom: Capture something far away

without moving.

Simply tap the number to switch—or pinch with two fingers to zoom manually.

## Focusing and Brightening

- **Tap** anywhere on the screen to focus on that spot.
- After tapping, slide your finger **up or down** to adjust brightness manually.

This is helpful when taking photos of people with backlighting or in very bright/dark areas.

## Using the Flash and Live Photos

- **Flash**: Tap the lightning bolt to turn flash on, off, or auto.
- **Live Photos**: This captures 1.5 seconds before and after your photo, creating a mini-video. Tap the icon of concentric circles to toggle it.

Live Photos are magical for capturing movement—like kids laughing or someone blowing out birthday candles.

## Editing Your Photos—Simply and Gently

After taking a photo, tap on it and choose "Edit."

You'll see options to:

- **Auto Enhance**: Automatically improve lighting and clarity

- **Crop or Rotate**
- **Adjust Brightness, Contrast, and Shadows**
- **Add Filters:** Make it warm, vivid, dramatic, or black & white

Don't be afraid to try things. Edits are non-destructive—you can always revert to the original.

## Photo Tips That Change Everything

1. **Clean the lens** with a soft cloth—it makes a bigger difference than you think.
2. **Use grid lines** (Settings > Camera > Grid) to help frame your shot.
3. **Get closer**, don't zoom, when taking photos of people or food.
4. **Use natural light** whenever possible—it's always more flattering.
5. **Take multiple shots** of a moment—then choose the best one later.

## Organizing and Finding Your Photos

Your iPhone organizes your pictures for you in the **Photos app.**

- **Recents:** All your newest photos and videos
- **Favorites:** Tap the heart ♥ icon to save ones you love

- **Albums:** Automatically sorted by people, places, screenshots, and more
- **Search:** Type "dog," "beach," or "July" to find specific moments

It's like having a digital memory book at your fingertips.

## Video Recording Made Easy

- Tap "Video" in the camera
- Tap the red button to start recording
- Tap it again to stop

Your iPhone automatically stabilizes the footage and adjusts brightness. For important moments (like weddings, birthdays, or grandkids dancing), this feature alone makes the phone worth it.

## You Are the Photographer Now

It doesn't matter if you've never owned a camera before. What matters is your eye, your story, your moment.

And now, you have the tool to capture it—with zero stress and beautiful results.

Take your time, experiment, and remember: your best photo isn't just about perfect lighting. It's about *what it means to you.*

# Chapter 6

## Keep Your iPhone (and Data) Safe

### Protecting Your Privacy Without Panic

Your iPhone 16 Pro isn't just a phone. It holds your contacts, your photos, your conversations, your appointments—and even your memories.

That's why **security isn't a luxury. It's a necessity.**
And with the iPhone, keeping your personal information safe is more doable than you might think.

In this chapter, we'll walk through the tools Apple provides to protect your identity, your device, and your peace of mind. You don't need to be a tech expert. You just need to follow a few simple steps—and you'll be in control, not your fears.

### Why Security Matters

Phones are like keys to your digital life. Without protection, someone could access:

- Your private messages
- Your bank and payment apps
- Your location

- Your photos and documents
- Your saved passwords

**But here's the good news:**
The iPhone 16 Pro is one of the most secure consumer devices ever made. You just need to know how to use its built-in protection.

## Step 1: Use a Strong Passcode

Your first and strongest defense is your **lock screen passcode.**

If you haven't set one yet or want to change it:

1. Go to **Settings > Face ID & Passcode**
2. Enter your current code (if prompted)
3. Tap **Change Passcode**
4. Choose a **six-digit or alphanumeric code** (longer is stronger)

> Tip: Avoid birthdays, repeated numbers, or obvious patterns like 123456.

## Step 2: Set Up Face ID

Face ID makes your phone easy to unlock—but hard for others to access.

1. Go to **Settings > Face ID & Passcode**
2. Tap **Set Up Face ID**

3. Follow the instructions to scan your face in a circle

Now, your phone can unlock securely with just a glance. It's fast, hands-free, and private. Only you can unlock it—*even if someone looks similar to you.*

## Step 3: Turn on Find My iPhone

If your phone is ever lost or stolen, **Find My iPhone** can help you:

- Locate it on a map
- Make it play a sound
- Remotely lock or erase it

To check it's on:

1. Go to **Settings > [Your Name] > Find My > Find My iPhone**
2. Make sure it's **ON**
3. Also turn on **Send Last Location**

> Tip: You can use another device (like an iPad or a friend's phone) to track it atiCloud.com/find

## Step 4: Manage App Permissions

Not every app needs access to your microphone, camera, contacts, or location.

To see and control which apps can access what:

1. Go to **Settings > Privacy & Security**
2. Tap categories like **Location Services**, **Camera**, or **Microphone**
3. Review and switch off any access that doesn't feel necessary

Example: A calculator app probably doesn't need to know your location.

This gives you full control over your digital footprint—**without deleting anything.**

## Step 5: Enable Two-Factor Authentication

This adds an extra layer of security to your Apple ID.

With **two-factor authentication**, even if someone guesses your password, they can't log in without your approval.

To turn it on:

1. Go to **Settings > [Your Name] > Password & Security**
2. Tap **Two-Factor Authentication**
3. Follow the setup instructions

Every time you sign in to a new device, you'll get a one-time code sent to a trusted number or device.

## Step 6: Protect Your Photos and Notes

Some memories are just for you. Thankfully, your iPhone allows you to **hide and lock private photos and notes.**

### To Hide Photos:

1. Open the **Photos** app
2. Tap **Select**, choose photos, then tap the **More (...)** icon
3. Choose **Hide**

They'll be moved to a secure, hidden album.

To lock them with Face ID:

1. Go to **Settings > Photos**
2. Toggle on **Use Face ID for Hidden and Recently Deleted**

### To Lock Notes:

1. Open the **Notes** app
2. Swipe left on the note title and tap the **lock icon**
3. Use Face ID or passcode to lock/unlock it

## Step 7: Secure Your Messages

iMessage is end-to-end encrypted by default—which means only the sender and receiver can read the messages.

To enhance privacy even more:

- Avoid sharing sensitive data in texts when possible
- Enable **Message Expiration:** Go to **Settings > Messages > Keep Messages** and set it to 30 days or 1 year
- Use **Mail Privacy Protection:** Go to **Settings > Mail > Privacy Protection** and turn it on to block tracking pixels

## Step 8: Use Safety Check (For Emergency Situations)

Apple includes a powerful feature called **Safety Check** for people in vulnerable or abusive situations.

If you ever need to quickly disconnect from someone who has access to your data, location, or device:

1. Go to **Settings > Privacy & Security > Safety Check**
2. Choose **Emergency Reset** or **Manage Sharing & Access**

This gives you instant control to stop location sharing, revoke access, and reset privacy settings.

## You Are Your Own Best Protection

Security can feel overwhelming at first, but your iPhone was built to protect you—*not confuse you.*

You don't need to be paranoid. Just proactive.

Take a few minutes today to walk through these steps. Even if you just do one or two, you're already safer than most people who never think about it at all.

# Chapter 7

## Battery & Storage Tips to Keep Your Phone Fast

### Make Your iPhone Last Longer, Run Smoother, and Feel Brand New

Your iPhone 16 Pro is powerful, but even the best technology can slow down if it's overloaded or not optimized.

You've probably asked yourself questions like

- "Why is my battery draining so fast?"
- "Why is my phone getting slower?"
- "How do I clear up space without losing anything important?"

This chapter will answer all of that and more.

You don't need to download extra apps, erase everything, or become a tech expert. With a few small habits and built-in settings, you can make your iPhone faster, lighter, and longer-lasting.

## Let's Start with the Battery: Why It Drains

There are several common reasons why your iPhone battery might run out quickly:

- Too many apps running in the background
- Bright screen settings
- Constant notifications
- Poor Wi-Fi or cellular signal
- Location services always on
- Background app refresh constantly running

Now, let's fix those things—without sacrificing what you love about your phone.

## Step 1: Turn on Low Power Mode (When Needed)

Low Power Mode reduces background activity and boosts battery life.

To activate:

1. Go to **Settings > Battery**
2. Toggle **Low Power Mode** ON

Or ask Siri, "Turn on Low Power Mode."

This is great when your battery dips below 30%, or you know you'll be out for the day without charging.

## Step 2: Adjust Screen Brightness and Auto-Lock

Your screen is one of the biggest battery users.

To reduce battery drain:

- Go to **Settings > Display & Brightness**
- Lower the **Brightness** slider
- Turn on **Auto-Brightness** if it's not already
- Set **Auto-Lock** to 30 seconds or 1 minute (under **Settings > Display & Brightness > Auto-Lock**)

## Step 3: Manage Background App Refresh

Some apps constantly update even when you're not using them. That can drain battery and data.

To disable:

1. Go to **Settings > General > Background App Refresh**
2. Choose **Off** or **Wi-Fi only**, or turn it off for individual apps

Turn it off completely unless you need real-time updates (like for weather or maps).

## Step 4: Disable Unnecessary Location Services

Apps don't always need to know where you are.

1. Go to **Settings > Privacy & Security > Location Services**
2. Review the list of apps
3. Set apps to **Never** or **While Using** as needed

It's good to disable location access for apps like social media or games unless location is essential.

## Step 5: Review Battery Usage by App

You can see exactly what's draining your battery.

1. Go to **Settings > Battery**
2. Scroll down to **Battery Usage by App**

If you see an app using a lot of power—even when you're not using it—consider deleting it, disabling background activity, or using it less.

## Now Let's Talk Storage: Keeping Your iPhone Clutter-Free

Storage fills up fast—especially with photos, videos, apps, and messages. But you don't need to delete things blindly. Let's manage it smartly.

## Step 1: Check Your Storage

1. Go to **Settings > General > iPhone Storage**

Here, you'll see:

- How much space is used
- What's using it (photos, apps, messages, etc.)
- Suggestions to optimize space

You'll also get helpful recommendations like "Offload Unused Apps" or "Review Large Attachments."

## Step 2: Offload Unused Apps (Without Losing Data)

1. Go to **Settings > General > iPhone Storage**
2. Tap an app you rarely use
3. Tap **Offload App**

This removes the app but keeps its documents and settings. If you reinstall later, it'll be just like you left it.

## Step 3: Delete Old Messages and Attachments

1. Go to **Settings > Messages > Keep Messages**
2. Change from **Forever** to **1 Year** or **30 Days**

This automatically deletes older texts and media you no longer need.

To manually delete large message attachments:

1. Go to **Settings > General > iPhone Storage > Messages**
2. Tap **Review Large Attachments**

3. Delete items like videos or voice notes you don't need

## Step 4: Clean Up Your Photos

Photos and videos take up the most space.

To free up room:

- Delete duplicates or blurry shots
- Empty the **Recently Deleted** album (in **Photos > Albums > Recently Deleted**)
- Consider using **iCloud Photos:**
    - Go to **Settings > [Your Name] > iCloud > Photos**
    - Turn on **iCloud Photos** and **Optimize iPhone Storage**

This stores full-resolution versions in the cloud and keeps smaller versions on your device.

## Step 5: Restart Your iPhone Once a Week

Just like your body, your phone benefits from rest.

Restarting clears memory, closes hidden background tasks, and refreshes performance.

To restart:

- Hold the **Power Button + Volume Up** until "slide to power off" appears

- Slide it, wait 10 seconds, then press Power again to restart

## Bonus Tip: Keep Your iPhone Updated

Software updates include performance improvements and bug fixes that often reduce battery drain and speed up performance.

To check for updates:

- Go to **Settings > General > Software Update**

Install updates when available, especially major iOS releases.

## Your iPhone Should Work *For* You, Not Slow You Down

When your phone runs smoothly, your life flows more easily.

And with a few smart habits, you don't need to be tech-savvy to keep your iPhone fast, clean, and fully charged.

# Chapter 8

## Hidden Features Most People Miss

### Delightful Tricks, Time-Saving Tools, and Everyday Superpowers

The iPhone 16 Pro is like a house with secret doors. On the surface, everything works great—but once you know where to look, you'll find shortcuts and hidden tools that make life easier, faster, and even a little fun.

In this chapter, we'll explore small features that make a **big difference**—especially for seniors, beginners, or anyone who wants to get more out of their iPhone without digging through every setting.

These aren't gimmicks. They're **genuinely useful** tools hiding in plain sight.

Let's unlock them.

### 1. Back Tap: Turn Taps into Shortcuts

Back Tap lets you double- or triple-tap the back of your iPhone to perform actions like

- Take a screenshot

- Open the flashlight
- Launch your camera
- Scroll up or down
- Open a favorite app

To activate:

1. Go to **Settings > Accessibility > Touch > Back Tap**
2. Choose **Double Tap** or **Triple Tap**
3. Assign it a function (like Screenshot or App Switcher)

Now, just gently tap the back of your phone and watch the magic happen.

## 2. Live Text: Copy Words From the Real World

You can point your camera at a document, sign, or handwritten note—and instantly copy the text.

How it works:

- Open the **Camera** and point it at text
- A yellow frame will appear around the words
- Tap the **Live Text** icon in the corner
- You can now **copy, share, translate, or look up** any text from the image

This works for phone numbers, addresses, prices, menus—anything. It's like having a scanner, translator, and secretary in your pocket.

## 3. Visual Look Up: Discover More in Your Photos

Your iPhone can identify objects in your pictures—like plants, landmarks, dog breeds, books, or artwork.

To use it:

1. Open a photo in the **Photos** app
2. If recognized, you'll see an **"i" with stars** icon—tap it
3. You'll get detailed info about what's in the image

This is great for curious minds, travelers, or anyone who's ever said, "What kind of flower is that?"

## 4. Picture-in-Picture: Multitask Like a Pro

Watching a video or FaceTiming and need to check a message? No problem.

Picture-in-picture lets you **minimize** the video into a floating window that stays on-screen while you do other things.

To activate:

- Start playing a video (in Safari, FaceTime, or supported apps)
- Swipe up from the bottom to go Home

- The video will shrink to a corner—drag it wherever you like
- Tap it to pause, resume, or return to full screen

Now you can watch and work at the same time—effortlessly.

## 5. Quick Notes: Jot Things Down Instantly

You can write a note from anywhere without opening the Notes app.

Here's how:

- Swipe your finger from the bottom-right corner of the screen (if enabled)
- A new Quick Note appears
- Type or dictate your thoughts
- Tap Done—it's saved to Notes automatically

You can enable this gesture in **Settings > Notes > Quick Note > Corner Gestures**

It's perfect for sudden ideas, grocery lists, or things to remember later.

## 6. Tap and Drag to Select Multiple Items

In apps like Messages, Mail, or Files, you can select many items at once—*faster than tapping them one by one.*

Try this:

1. Touch and hold an item (like a message or photo)
2. With another finger, tap additional items to add them to the group
3. Move or delete them together

It's a small thing, but it makes a big difference when cleaning up your phone or sharing multiple files.

## 7. Use Siri Without Talking (Type to Siri)

Prefer to type instead of speaking? You can still use Siri silently.

To turn this on:

1. Go to **Settings > Accessibility > Siri**
2. Turn on **Type to Siri**

Now, when you activate Siri, a keyboard appears instead of a voice prompt. You can type:

- "Set a timer for 15 minutes."
- "Text John, I'm on my way."
- "What's the weather today?"

Quiet, private, and just as helpful.

## 8. Drag and Drop Between Apps

You can now drag content—like text, images, or files—from one app to another.

Here's how:

1. Tap and hold a photo (or selected text)
2. With another finger, swipe up to open another app
3. Drop the item where you want (like into a message or document)

This works in **Photos, Messages, Mail, Notes, Safari**, and many other apps.

## 9. Silence Unknown Callers

Tired of spam and robocalls? You can stop unknown numbers from ringing altogether.

To enable:

- Go to **Settings > Phone > Silence Unknown Callers**
- Turn it ON

Now only people in your contacts will ring through. Unknown numbers will still show in your Recents, but they won't disturb you.

## 10. Handoff: Start on One Device, Finish on Another

If you use other Apple devices (like an iPad or Mac), you can **start something on one device and continue on another.**

- Start writing an email on your iPhone
- Open your iPad or Mac—it'll show up at the bottom to

continue

Make sure Bluetooth is on and you're signed in to the same Apple ID across devices. You'll find the setting under **Settings > General > AirPlay & Handoff**

## Small Tricks That Make a Big Difference

Here are a few more hidden treasures worth trying:

- **Shake to Undo**: Made a mistake while typing? Shake the phone gently to undo.
- **One-Handed Keyboard**: Tap and hold the keyboard icon to shift keys left or right for one-handed use.
- **Create Custom Vibration Alerts**: Go to **Settings > Sounds & Haptics > Ringtone > Vibration > Create New Vibration**
- **Pin Conversations in Messages**: Swipe right on any message thread to pin it at the top.

## You Know More Than Most iPhone Users Now

Most people never explore these features. But now *you* have.

You've taken your iPhone beyond the basics and turned it into a personal assistant, creative tool, and time-saver.

# Chapter 9

## Hands-Free Living with Siri & Automation

### Let Your iPhone Do the Work, So You Don't Have To

Imagine if your iPhone could do things *before* you even think of them.

- Set reminders while you're driving
- Text someone while your hands are full
- Turn off your lights when you leave the house
- Automatically silence notifications when you're sleeping

This isn't futuristic fiction—it's your iPhone 16 Pro working smarter for you.

In this chapter, we'll explore how to live more hands-free using **Siri** and **Shortcuts**—Apple's powerful yet easy-to-use automation system. You'll learn how to give simple voice commands, automate daily routines, and turn your iPhone into a true personal assistant.

No coding, no complexity—just real, practical ways to do more with less effort.

## What Is Siri?

Siri is Apple's built-in voice assistant. It listens, understands, and responds to your voice.

You can use Siri to:

- Make calls
- Send texts
- Set alarms, timers, and reminders
- Play music or podcasts
- Get directions
- Control smart home devices
- Ask questions (weather, sports, trivia, etc.)

It's like having a smart helper in your pocket—all you have to do is ask.

## How to Activate Siri

There are a few ways to bring Siri to life:

- **Voice Activation**: Say "Hey Siri" (or just "Siri" on newer models)
- **Button Activation**: Press and hold the **Side Button** (on the right side of your iPhone)
- **Type to Siri**: If you prefer to type, enable this in **Settings**

**> Accessibility > Siri > Type to Siri**

Make sure Siri is enabled by going to

**Settings > Siri & Search**
Turn ON:

- **Listen for "Hey Siri."**
- **Press Side Button for Siri**
- **Allow Siri When Locked**

Now Siri is ready—whenever you are.

## Practical Voice Commands You Can Use Today

Here are everyday commands to help you get started:

**For Communication**

- "Siri, call Linda."
- "Siri, text Emma 'I'm on my way.'"
- "Siri, check my voicemail."

**For Productivity**

- "Siri, remind me to take my pills at 9 PM."
- "Siri, set a timer for 20 minutes."
- "Siri, what's on my calendar today?"

**For Entertainment**

- "Siri, play jazz music."
- "Siri, skip this song."
- "Siri, what's the name of this artist?"

**For Navigation**

- "Siri, take me to the nearest pharmacy."
- "Siri, what's the traffic like to work?"
- "Siri, how long will it take to get to the airport?"

**For Weather and Information**

- "Siri, what's the weather tomorrow?"
- "Siri, how old is Tom Hanks?"
- "Siri, define 'serendipity.'"

## Making Siri Work for You Personally

Siri learns from your usage and preferences. To make it even smarter:

- Add **relationships** in your Contacts
  - Example: Edit your spouse's contact and label them as "spouse." Then say, "Call my spouse."
- Teach Siri **pronunciations** by correcting mispronounced names
- Allow Siri access to your **apps** in **Settings > Siri & Search**

The more you use Siri, the more useful it becomes.

## What Are Shortcuts?

**Shortcuts** are custom commands that let your iPhone perform multiple tasks in a single tap or voice request.

Think of them as personal routines. You can:

- Send a "Running Late" message to your boss
- Set your phone to Do Not Disturb, lower brightness, and play calming music
- Launch directions to your favorite café and start a podcast—all in one tap

You can trigger Shortcuts with Siri or as automations that run based on time, location, or activity.

## How to Use the Shortcuts App

The Shortcuts app is built into your iPhone 16 Pro. Here's how to use it:

1. Open **Shortcuts**
2. Tap **Gallery** to browse pre-made shortcuts
3. Tap **Add Shortcut** to install one
4. Tap the shortcut to run it—or say its name to Siri

Examples:

- "Heading" Home"—Sends ETA, opens Maps, starts your playlist
- "Laundry" Sets a 40-minute timer
- "Water Tracker"—Adds" a glass of water to your Health app

You can also build your own, but don't worry about that unless you feel adventurous. Start with the ready-made ones—they're excellent.

## Automating Your Life with Personal Automations

Want your phone to do something *automatically* based on time, location, or action?

Try this:

1. Open **Shortcuts > Automation > Create Personal Automation**
2. Choose a **trigger** (Time of Day, Arrive at Location, App Opened, etc.)
3. Set the **action** (Send message, turn on Focus Mode, play music, etc.)

Examples:

- Automatically turn on Do Not Disturb at bedtime
- Lower screen brightness when battery drops below 20%
- Launch your workout playlist when you open the Fitness

app

You're not just using your phone—you're choreographing your day.

## Hands-Free for Accessibility and Peace of Mind

Siri and automation aren't just about convenience. They also support:

- **Mobility limitations** (hands-free commands)
- **Memory challenges** (scheduled reminders and alerts)
- **Safety** (emergency contacts, "Hey Siri, call 911")
- **Smart home control** (lights, thermostat, locks)

With Siri, even a moment when your hands are full, eyes are busy, or you're feeling overwhelmed becomes manageable.

## You've Earned the Right to Let Your iPhone Help You

Technology should work for you, not wear you out.

By learning to use Siri and automation, you're simplifying your life, reducing daily friction, and turning your iPhone into a reliable helper.

Even if you only use one or two voice commands or shortcuts, you've already taken a step toward less stress and more freedom.

# Chapter 10

## Top 20 Must-Have Apps in 2025

### The Best Apps for a Simpler, Smarter, and More Enjoyable iPhone Experience

Your iPhone is like a toolbox—and apps are the tools you choose to keep inside.

But with millions of apps out there, how do you know which ones are *really worth your time*?

In this chapter, we've carefully selected 20 essential apps that will make your iPhone more useful, more enjoyable, and better suited to your unique lifestyle. Whether you're a senior exploring technology, a beginner learning the ropes, or someone looking for practical solutions, these apps were chosen with *you* in mind.

You don't need all of them. Just pick the ones that match your needs—and watch your phone become more helpful every day.

### Productivity & Organization

1. **Google Calendar**
   An easy-to-use calendar that syncs with all your devices. Schedule appointments, set reminders, and never miss a date.

2. **Evernote**

   A powerful note-taking app that lets you write, record voice memos, and clip web articles—all in one place.

3. **Reminders (Apple)**

   Already on your phone—use it to set shopping lists, medication alerts, and household to-dos.

4. **Files (Apple)**

   Organize your documents, PDFs, and downloads. Think of it like a digital filing cabinet built right into your iPhone.

## Health, Wellness & Safety

1. **Apple Health**

   Tracks your steps, medications, heart rate, sleep, and more. Connects with other apps like your smart scale or fitness watch.

2. **Medisafe**

   Never miss a dose again—this medication reminder app is easy, clear, and especially helpful for managing multiple prescriptions.

3. **Magnifying Glass + Flashlight**

   Turns your iPhone into a magnifier for reading small print—menus, medication labels, or fine print on documents.

4. **Find My (Apple)**

   Already on your phone. Use it to locate your iPhone, AirPods, or even a lost loved one (with permission) in seconds.

## Learning & Reading

1. **Kindle**
   Turn your iPhone into a portable library. Download books, adjust font size, and read comfortably anywhere.

2. **Libby**
   Borrow ebooks and audiobooks for free from your local library—right on your phone.

3. **YouTube**
   Learn anything from cooking to gardening, language skills to tech tutorials. It's like the world's largest classroom.

4. **Duolingo**
   A fun, friendly way to learn a new language even 5 minutes a day can expand your world.

## Communication & Connection

1. **WhatsApp**
   Make free calls, video chats, and group messages—especially useful for international family and friends.

2. **Zoom**
   For face-to-face calls with kids, grandkids, friends, or meetings. Simple to use and widely accepted.

3. **Telegram**

A secure and private messaging app with excellent group chat features. Fast, lightweight, and ad-free.

## Creativity & Fun

1. **Canva**
   Design birthday cards, flyers, social media posts, or even photo collages—no experience needed.

2. **Apple Photos + Editing Tools**
   Already on your iPhone—use it to enhance images, create albums, and make memories look even better.

3. **Spotify**
   Stream music, podcasts, or calming playlists for walking, relaxing, or dancing around the kitchen.

4. **Pinterest**
   Explore ideas for home décor, recipes, crafts, fashion, and more. It's a visual inspiration board at your fingertips.

## Bonus Pick: Utility That Changes Everything

1. **1Password** or **Apple Keychain (built-in)**
   Keep track of all your passwords in one secure place—so you never have to remember them again. Safer than writing them down.

## A Few Installation Tips

- All apps can be downloaded from the **App Store**.

- To download: Open **App Store > Search > Type the app's name > Tap "Get" or the price.**
- Use your **Face ID or Apple ID password** to confirm.

If you're concerned about space, don't install them all at once. Try one or two that feel useful now. You can always add more later.

## You Don't Need Every App—Just the Right Ones

Apps should **serve you**, not overwhelm you. With the right ones, your iPhone becomes:

- A reliable assistant
- A source of joy and entertainment
- A lifeline to family and friends
- A gateway to learning new skills
- A tool for better health and security

Choose apps that bring ease, clarity, and enjoyment to your life—and let the rest go.

# Chapter 11

## Troubleshooting & Updates

### Solving Common iPhone Problems Without the Panic

Even the best devices have their off days.

Sometimes your iPhone may freeze, a message won't send, or an app won't open. These moments can feel frustrating—but they're almost always easy to fix.

In this chapter, you'll learn how to handle the most common iPhone issues calmly and quickly. No tech support needed, no trips to the Apple Store. Just you, your phone, and a few smart tricks.

We'll also show you how to check for software updates and install them properly, because staying up-to-date helps prevent problems before they start.

### Part 1: Common Problems & Easy Fixes

#### Problem: The Screen Is Frozen or Unresponsive

**Solution**: Force Restart

1. Quickly press and release the **Volume Up** button

2. Quickly press and release the **Volume Down** button
3. Press and hold the **Side Button** (on the right) until the Apple logo appears

This doesn't erase anything. It just restarts the phone.

## Problem: No Internet Connection (Wi-Fi or Cellular)

**Solution:**

- Check **Settings > Wi-Fi** to make sure you're connected
- If using cellular, go to **Settings > Cellular** and ensure it's turned on
- Turn **Airplane Mode ON**, wait 5 seconds, then turn it **OFF**
- Still stuck? Restart the iPhone

## Problem: Messages Aren't Sending or Receiving

**Solution:**

- Make sure you have a network connection (Wi-Fi or cellular)
- Go to **Settings > Messages**, then turn **iMessage OFF**, wait, and turn it **ON** again
- Restart the phone
- Ask the other person if their phone is working—sometimes

it's on their end

## Problem: An App Keeps Crashing

**Solution:**

1. Swipe up from the bottom and pause to open the App Switcher
2. Swipe the app off the screen to close it
3. Reopen the app
4. If that fails:
    - Go to the **App Store**
    - Tap your **profile icon** in the top-right
    - Scroll down to check for **App Updates**
    - Update the app if available

If nothing works, delete the app and reinstall it from the App Store.

## Problem: Battery Is Draining Too Fast

**Solution:**

- Turn on **Low Power Mode** (Settings > Battery)
- Check **Battery Usage by App** to see which apps are draining power
- Lower screen brightness and reduce background app

refresh

- Update your software (see next section)

## Problem: No Sound or Sound Is Muffled

**Solution:**

- Make sure the **Mute switch** on the left side of your iPhone is OFF
- Check the **Volume** using the side buttons
- Go to **Settings > Sounds & Haptics** and increase the volume
- Restart the phone
- For muffled audio, clean the speaker openings gently with a dry toothbrush or soft cloth

## Problem: Camera Isn't Working

**Solution:**

- Close the Camera app and reopen it
- Make sure no other app (like FaceTime) is using the camera
- Restart the phone
- Check for any screen protector or case blocking the lens

## Part 2: How to Update Your iPhone Software

Keeping your iPhone updated helps.

- Fix bugs
- Improve performance
- Add new features
- Patch security vulnerabilities

And best of all? It's easy.

### How to Check for Updates

1. Go to **Settings > General > Software Update**
2. If an update is available, you'll see **Download and Install**
3. Tap it, agree to the terms, and wait while your iPhone updates

> Tip: Connect to Wi-Fi and plug your phone into a charger before updating.

Updates can take a few minutes to an hour, depending on your internet speed and phone model. Don't worry—your data stays safe.

## Part 3: When to Ask for Help

Sometimes, it's okay to get support—especially if:

- You've tried the solutions and the issue continues

- The screen is physically damaged or cracked
- Your iPhone won't turn on at all
- You're locked out and can't remember your Apple ID or password

In those cases, visit **support.apple.com** or use the **Apple Support** app (free from the App Store).

You can also call Apple Support or visit an Apple Store or Authorized Service Provider in your area.

## You're More Capable Than You Think

Every time you solve a problem on your own, you're proving something:
**You're not behind. You're learning. You're growing. And you're mastering your iPhone.**

Troubleshooting doesn't have to be scary. With calm, clear steps, you now have the power to handle most problems in minutes.

# Chapter 12

## Quick Reference Cheat Sheets

### Your At-a-Glance Guide to Everyday iPhone Success

Sometimes, you just want a quick answer—without digging through menus or searching online.

That's exactly what this chapter is for.

Whether you need to remember how to take a screenshot, ask Siri for something, or change a setting, these cheat sheets give you fast access to what matters most.

You can even print them out or screenshot them for easy access anytime.

### Navigation & Gestures

- **Go to Home Screen:** Swipe up from the bottom
- **See Open Apps:** Swipe up and hold
- **Access Control Center:** Swipe down from top-right corner
- **See Notifications:** Swipe down from top-left or center

- **Search Your iPhone**: Swipe down on Home Screen
- **Go Back**: Swipe from the left edge to the right
- **Close App**: Swipe up from bottom and swipe the app away

## Battery & Power

- **Low Power Mode:**
  Settings > Battery > Toggle ON
- **Check Battery Usage:**
  Settings > Battery > See usage by app
- **Force Restart:**

  a. Press Volume Up

  b. Press Volume Down

  c. Hold Side Button until Apple logo appears

## Camera & Photos

- **Open Camera Fast:**
  Swipe left from Lock Screen
  Or use the Camera app
- **Take a Screenshot:**
  Press Side Button + Volume Up
- **Record Video:**
  Swipe to Video in Camera app, tap red circle

- **Use Live Text:**
  Open Camera, point at text, tap yellow frame icon
- **Hide Photos:**
  Photos > Select > More (...) > Hide
- **Favorite a Photo:**
  Tap the □ icon on any photo

## Siri Voice Commands

- **"Hey Siri, call [Name]."**
- **"Hey Siri, text [Name], 'I'll be there soon.'"**
- **"Hey Siri, what's the weather today?"**
- **"Hey Siri, set a timer for 10 minutes."**
- **"Hey Siri, remind me to water the plants at 6 PM."**
- **"Hey Siri, open Camera."**

To turn Siri on:
Settings > Siri & Search > Enable "Hey Siri"

## Security & Privacy

- **Set or Change Passcode:**
  Settings > Face ID & Passcode
- **Enable Face ID:**
  Settings > Face ID & Passcode > Set Up Face ID

- **Turn on Find My iPhone:**
  Settings > [Name of your device] > Find My > Find My iPhone

- **Manage App Permissions:**
  Settings > Privacy & Security

## Notifications & Sounds

- **Silence Unknown Callers:**
  Settings > Phone > Silence Unknown Callers

- **Change Ringtone:**
  Settings > Sounds & Haptics > Ringtone

- **Do Not Disturb / Focus Mode:**
  Settings > Focus > Choose Focus
  Or swipe down top-right > tap Focus icon

## Storage & Cleanup

- **Check Storage:**
  Settings > General > iPhone Storage

- **Offload Unused Apps:**
  Settings > iPhone Storage > Tap app > Offload

- **Delete Message Attachments:**
  Settings > iPhone Storage > Messages > Review Attachments

- **Clear Safari History:**
  Settings > Safari > Clear History and Website Data

## Software Updates

- **Check for Updates:**
  Settings > General > Software Update
  Tap "Download and Install" if available

## Message Tips

- **Pin a Conversation:**
  Swipe right on a message thread > Tap Pin

- **Reply to a Specific Message:**
  Press and hold on a message > Tap "Reply."

- **React to a Message:**
  Tap and hold > Choose emoji reaction

## Pro Tips Worth Remembering

- **Back Tap:**
  Settings > Accessibility > Touch > Back Tap
  Set double or triple tap to do things like screenshot or open flashlight

- **Quick Note Swipe:**
  Swipe from bottom-right corner to write a new note instantly

- **Picture-in-Picture:**
  Play a video > Swipe up to keep watching while using other apps

- **Search App Library:**
  Swipe all the way right > Type app name in search bar

## You Now Hold the Knowledge

These cheat sheets aren't just shortcuts—they're proof of how far you've come.

From powering up your phone for the first time to mastering features that most users never discover, you've taken real steps to understand, enjoy, and own your iPhone 16 Pro.

And the best part? You can always return to this chapter if you forget something or want a quick reminder.

If you're ready for a little bonus, the next section will help you go even further: resources for learning more, getting support, and keeping your iPhone experience joyful, secure, and stress-free.

# About the Author

**Blake Orion** is a bestselling technology educator, digital accessibility advocate, and author of over a dozen beginner-friendly tech guides designed for seniors, caregivers, and everyday users who want to feel more confident with modern technology—without the overwhelm.

With a background in digital learning and user experience design, Blake has spent more than 15 years helping thousands of people across all ages bridge the gap between frustration and freedom with their devices. His step-by-step manuals are trusted for their clear explanations, large-print formatting, and practical, no-jargon guidance that makes technology approachable, even for first-time users.

Blake's mission is simple: **make technology work for everyone—especially those left behind in the digital age.**

Made in the USA
Monee, IL
04 November 2026

67384059R00046